FOSS Science Resources

Pebbles, Sand, and Silt

Full Option Science System
Developed at
The Lawrence Hall of Science,
University of California, Berkeley
Published and distributed by
Delta Education,
a member of the School Specialty Family

1487700
978-1-62571-298-1
Printing 6 — 11/2017
Standard Printing, Canton, OH

Table of Contents

Exploring Rocks

Think about a **rock**. A rock has many **properties**. What does the rock look like?

Rocks can be small or large. They can be heavy or light. They can be smooth or rough. They can be round or flat, shiny or dull. Rocks can be different in many ways.

Some rocks are too big to hold in
your hand. A rock can be as big
as a mountain!

Other rocks are so small that you can hold thousands in your hand. Look at the picture of a **sand dune**. Can you see the tiny rocks blowing in the **wind**?

Rocks of all sizes can be found in rivers. Over time, rocks in a river become smooth. Rocks become smooth from rubbing against one another.

Rocks of all sizes can be found in a desert, too.
How big is the rock you're thinking of?

Rocks can be many different colors. They can be black, brown, red, or white. They might even be pink or green. Some rocks have speckles or stripes, too.

Rocks can be many different sizes. They can have different **textures**. They can be many colors and shapes. They can even have patterns.

What does the rock you're thinking about look like?

Colorful Rocks

What are these colorful objects?

They are **minerals**. There are many different kinds of minerals. Minerals come in lots of different colors.

Rocks are made out of minerals.
That's why rocks can be so many
different colors.

This rock is made of different minerals.
Can you see them?

Look for the black mineral. Look for the
pink mineral. Look for the gray mineral.
These are the minerals in this rock. This
rock is called pink **granite**.

The Story of Sand

Have you ever looked at one grain of **sand** and thought, "I wonder how it got so small?"

A grain of sand wasn't always so small! It might have once been part of a **boulder**. The boulder could have broken off a mountain. The boulder could have tumbled down the mountain.

Maybe the boulder rolled into a river. Water in a river can move rocks. The rocks bump together in the water. The boulder might have broken into **cobbles** and **pebbles**. Cobbles are bigger than pebbles.

Maybe the river carried the pebbles to the ocean. Ocean waves crash over pebbles. The pebbles might have broken into **gravel**. Pebbles are bigger than gravel.

Wind and water move and rub rocks together. Over time, rocks break apart. They can get smaller and smaller. They can break into very tiny rocks. This is called **weathering**. These tiny rocks are grains of sand.

Compare the sand from different places.

Corpus Christi, Texas

Dawson City, Texas

Green Sands Bay, Hawaii

Plum Island, Massachusetts

**Cape Hatteras National
Seashore, North Carolina**

Dix Beach, North Carolina

Next time you build a sand castle, think about
the story of sand!

Thinking about The Story of Sand

1. Put these rocks in order by size, from the largest to the smallest.

       sand

       boulder

       gravel

       cobble

       pebble

2. Tell the story of sand.

Rocks Move

Water and wind move rocks of all sizes.

Look at the pictures on these two pages. Can you tell what moved the rocks?

Mudflat

Sandy Beach

Washout

Landforms

Some landforms are formed by eruption.

A **volcano** is a place where lava, ash, and **gases** escape from openings in Earth's crust.

A cinder cone is a kind of volcano. It forms when cinders (pieces of lava) burst out of Earth in an eruption.

A shield volcano forms from flowing lava that has cooled. It is wider than it is tall.

A composite volcano forms from different kinds of eruptions. Layers of cinders, lava, and ash build up into a mountain.

Some landforms are formed by weathering and erosion.

A **valley** is a low area between mountains. Rivers make V-shaped valleys. Glaciers make U-shaped valleys.

A **canyon** is a deep V-shaped valley. Rivers erode the steep sides of canyons.

A **mesa** is a wide, flat-topped hill. It has at least one steep side.

A **butte** is a hill with steep sides and a flat top.
A butte is smaller than a mesa.

A **beach** is an area at the edge of a lake or the ocean. Beaches are often sandy.

A **delta** is a fan-shaped deposit of earth materials. Deltas form at the mouth of a river or stream.

A **plain** is low, flat, level land. Plains are found in the center of the United States.

A sand dune is a mound or hill of sand formed by wind.

Some landforms rise above Earth's surface. They weather to create other landforms.

A **mountain** is a high, steep area of land. Some mountains are volcanoes, but not this one.

A **plateau** is an area of high, nearly flat land. Plateaus cover a lot of area.

Making Things with Rocks

People use rocks to make things. A quarry is a place where people dig rocks out of the ground.

Big pieces of rock are used to make big things.
Statues and churches are often made from rock.
People make things out of rock because it lasts
a long time.

Pebbles and gravel are part of the mixture called **asphalt**. Asphalt is used to pave streets and playgrounds.

Gravel and sand are used to make sidewalks.
The gravel and sand are mixed together with
cement and water. Cement is like glue. It
holds the mixture together. When the mixture
gets hard, it makes **concrete**.

Even the tiniest rocks are useful. **Clay** is made up of rocks that are tinier than sand! They are so small that you can't see only one rock with your eyes. People mold clay into many shapes.

Clay is used to make bricks. Bricks are used to make walls and buildings. Bricks are used to make walking paths, too.

The bricks are held together with concrete **mortar**. Look at the mortar between the bricks. It is made of cement and sand.

Some bricks are made of concrete. They are called cinder blocks. How are cinder blocks and bricks the same? How are they different?

Whatever their size, rocks are useful. People make lasting and beautiful things from rocks.

What Are Natural Resources?

Rocks are **natural resources**. Rock walls can be formed by nature. Rock walls can be made by people, too.

Look at the rock walls. Which ones are natural? Which ones are made by people?

Stepping stones and walking paths can be natural. Stepping stones and walking paths can be made by people, too.

Look at the walking paths. Which ones are natural? Which ones are made by people?

Rock gardens can be natural. Rock gardens can be made by people, too.

Look at the rock gardens. Which ones are natural? Which ones are made by people?

What Is in Soil?

Rocks are all around you. The **soil** under your feet has rocks in it. Some of the tiny rocks and minerals in soil are called **silt**. Silt is smaller than sand, but bigger than clay. Sand, clay, gravel, and pebbles can be in soil, too.

When plants and animals die, they become part of the soil. Plants and animals **decay** into tiny pieces called **humus**. Humus provides **nutrients** for plants. It also helps the soil **retain** water.

What is this animal that lives in soil? A worm! Worms are good for soil. They burrow through the soil. They break it apart and enrich the humus. Worms help plants grow by mixing and turning the soil.

Not all soil is alike. Some soil has more humus. Some has more clay or sand. Some has more pebbles and gravel. What differences do you see in these soils?

Testing Soil

Do plants grow better in soil or sand? Here's what you can do to find out.

1. Get four cups that are all the same size.

2. Fill two cups with potting soil that has lots of humus. Fill the other two cups with sand.

3. Plant three sunflower seeds in each cup.

4. Put the same amount of water in each cup.

5. Keep the cups in a sunny window, and record what happens.

Thinking about Testing Soil

1. Is this a good way to test the question?

2. Two students planted seeds in soil and sand. Look at the plants above. Which seeds grew better? Why do you think that happened?

3. Do the test yourself. Draw or write about your results.

Where Is Water Found?

Water is found everywhere on Earth. Water is part of every living thing. Every plant and animal is made of water. Even you are mostly made of water!

Fresh water is found in streams and rivers. Streams can be small like a creek. Rivers are larger streams of water.

Streams and rivers bring water into and out of ponds and lakes.

Fresh water is found in ponds and lakes, too. Ponds are small bodies of water. Lakes are larger and deeper bodies of water.

The water moves slowly in ponds and lakes. Sand and silt settle to the bottom of ponds and lakes.

Fresh water is our most important natural resource. Plants and animals need water to live and grow.

People use water to drink, cook, and wash.
People use water to grow food and to power
factories, too.

Most of the water on Earth is **salt water**.
Salt water is found in seas and the ocean.
The ocean is the largest body of salt water.
Seas are smaller than the ocean.

Salt water is found in salt marshes. They are muddy places next to seas. Salt marshes have lots of grasses and small plants. Salt marshes have slow-moving water.

Salt water is found in mangrove forests.
They are like salt marshes, but they have
trees and bushes. The roots of mangrove
trees help protect the shore.

Salt water is found in coral reefs. Coral reefs grow in warm, shallow seas. Coral reefs are made from corals. Corals are the hard parts of sea animals.

Salt water is found on sandy beaches and rocky shores, too. You can see the ocean water move back and forth in waves on beaches and shores.

Where is fresh water found in your community?

Where is salt water found in your community?

States of Water

Liquid water is one state of water. We can pour it into a glass to drink. We spray it from a hose to water plants. Liquid water can drip from a fountain.

We see liquid water as dewdrops in the morning.
We see it as rain falling to Earth, too.

Solid ice is another state of water. When it gets cold, water freezes into a solid. We can pack snow to make a snowball. We can catch a snowflake.

We can skate on ice. We can float
ice cubes in lemonade.

Gas is another state of water. We cannot
see water when it is a gas. But it is in the
air all around us. When the gas becomes
a liquid, we see it as a cloud or rain.

When the liquid becomes a solid, we see water as snow or hail.

Water is found in all three states on Earth.
Water can be a solid, a liquid, and a gas.

Can you see all three states of water here?

Erosion

What happened to this road? People once drove on this road. During a big storm, waves crashed against the shore. They washed away the soil under the road. Parts of the road were destroyed.

Waves are moving water that cause **erosion** on a coast. Ocean waves often erode the shore during storms. Coastal erosion can damage roads and buildings. Waves can also wash away all the sand on a beach.

Where else do we find moving water? Water flows downhill in rivers and streams. Heavy rain causes the water in streams and rivers to flood their banks. Fast flowing water erodes the banks.

Engineers designed a strong barrier for the bank of this river. The barrier will protect the road and buildings from erosion caused by flowing water.

It will block rainwater from the road. It will stop the river from flooding the road.

Look for different ways that people protect the edges of waterways from erosion. Some design solutions use heavy objects to cover soil and hold it in place.

Sandbags

Concrete shapes

Concrete bricks

Cattails

Bundles of sticks

Bundles of sticks can protect riverbanks from moving water. Another good way to protect banks against erosion is to grow water plants, like cattails, along the banks.

Moving water is not the only force that causes erosion. Moving air can cause erosion, too. Powerful winds can remove topsoil from a farm. The winds lift the valuable topsoil into the air and carry it far away.

Beaches can be eroded by wind, too. Strong winds that blow across a beach can lift sand onto nearby land.

People have learned how to slow down erosion of topsoil and beach sand. Farmers plant rows of trees or shrubs to block the wind near their fields.

People put low wooden fences on the beach to slow the blowing sand. The wind piles the sand up by the fences and makes dunes. Once dunes form high on the beach, beach grass can grow. The grass on the dunes slows erosion even more.

These two roads have problems. What problems can you see? What might have caused the erosion? What would you suggest as a solution for the problem?

Ways to Represent Land and Water

How would you describe your classroom to your grandparents? You could tell them it has eight tables, each with four chairs. You could tell them what the furniture looks like. You could show them a photograph, but a photograph might not show everything.

You could draw a picture to represent the classroom. The drawing might show the size, design, and location of some of the furniture.

= Chair

You could also provide a **map** of the classroom. A map could show where all the tables and chairs are in the room. A map is a view from directly overhead.

These ways to represent a room are also good ways to represent Earth's surface. You can use photographs, drawings, and maps to show the location, size, and kinds of land and water in an area.

This photograph shows Crater Lake in Oregon.
It is the deepest lake in the United States,
594 meters deep.

A drawing is a different way to show Crater Lake.

A map is a different way to represent Crater Lake. The map shows the shape of the lake. It shows roads and other nearby features surrounding the lake.

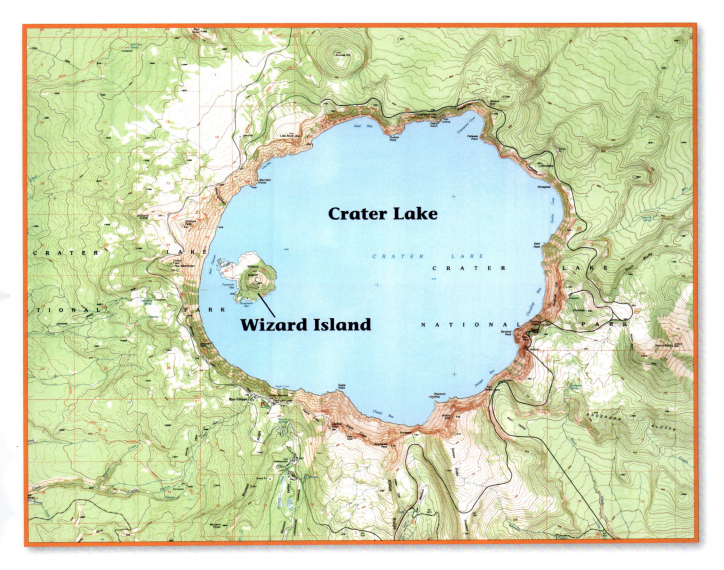

Crater Lake

Wizard Island

This photograph shows Mount Shasta. Mount Shasta is a volcano in Northern California. It is very tall, 4,322 meters.

A drawing is another way to show Mount Shasta.

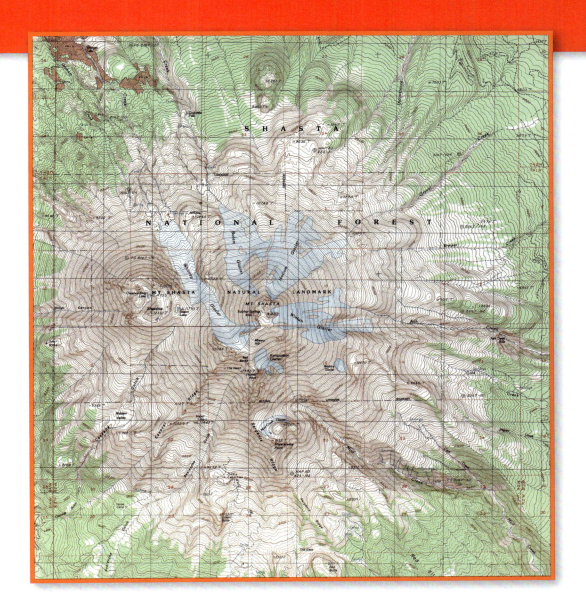

This map shows a topographic view of Mount Shasta. The lines show how high the land is. You can think of them as steps. The steps go up to the very top of the mountain. Water, streams, and ice appear in blue. Areas with trees are green. Maps can show a lot of information.

This photograph shows the Scioto River flowing through the city of Columbus, Ohio. This part of the river bank has a park. The park is called Scioto Mile.

This drawing is a different way to show the park.

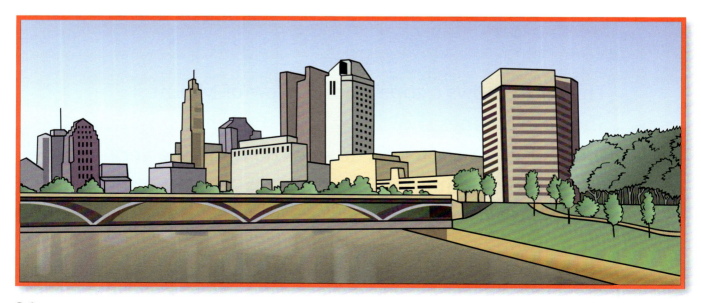

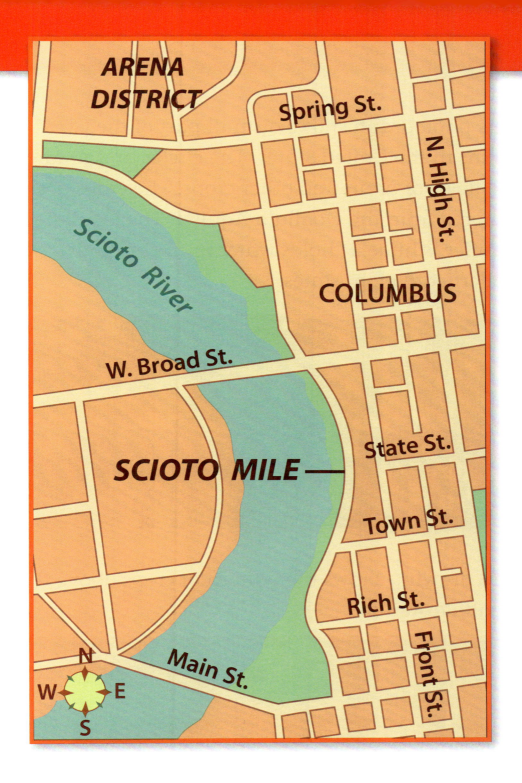

A map of the park shows how big the park is and where it is in Columbus.

This photograph shows a small part of the Great Plains in the United States. The circles and squares are fields of grain and other crops. The crops are watered from the Ogallala Aquifer. The Ogallala Aquifer is huge. It holds water in the ground underneath eight states.

This drawing shows the pattern of crops above the Ogallala Aquifer.

OGALLALA AQUIFER

Wyoming

South Dakota

Nebraska

Colorado

Kansas

New Mexico

Oklahoma

Texas

Ogallala Aquifer

0 50 100   200 MILES

SCALE

This map shows how big the aquifer is. It is located under parts of Oklahoma, Texas, New Mexico, Colorado, Kansas, Nebraska, Wyoming, and South Dakota.

This image of Earth shows the location of North America. Can you find the United States on the North American continent?

Here is a drawing showing the outline of the United States.

This map shows part of the United States, outlining 48 states. It also shows highways that connect the states, and large waterways (lakes, rivers, and the ocean).

Glossary

asphalt a mixture of pebbles and gravel **(33)**

beach an area at the edge of a lake or the ocean **(28)**

boulder a very large rock that is bigger than a cobble **(15)**

butte a hill with steep sides and flat top **(27)**

canyon a deep V-shaped valley **(26)**

cement a finely ground powder that is like glue when mixed with water **(34)**

clay rocks that are smaller than sand and silt. It is hard to see just one. **(35)**

cobble a rock that is smaller than a boulder, but bigger than a pebble **(16)**

concrete a mixture of gravel, sand, cement, and water **(34)**

decay when dead plants or animals break down into small pieces **(45)**

delta earth materials built up in the shape of a fan **(28)**

erosion the carrying away of earth materials by water, wind, or ice **(69)**

fresh water water without salt. Fresh water is found in streams, lakes, and rivers. **(51)**

gas matter that can't be seen but is all around. Air is an example of a gas. **(24)**

granite the name of a kind of rock. Pink granite is made of four minerals. Those minerals are hornblende (black), mica (black), feldspar (pink), and quartz (gray). **(13)**

gravel a rock that is smaller than a pebble, but bigger than sand **(17)**

humus bits of dead plant and animal parts in the soil **(45)**

liquid matter that flows freely and takes the shape of its container **(61)**

map a picture that shows different parts of something **(81)**

mesa a hill with a wide, flat top **(27)**

mineral the colorful ingredient that makes up rocks **(11)**

mortar a mixture of cement and sand **(36)**

mountain a high and steep area of land **(30)**

natural resource something from Earth. Rocks, soil, air, and water are natural resources. **(38)**

nutrient something that living things need to grow and stay healthy **(45)**

pebble a rock that is smaller than a cobble, but bigger than gravel **(16)**

plain a low, flat, level area of land **(29)**

plateau a high, nearly flat area of land **(30)**

property something that you can observe about an object or a material. Size, color, shape, texture, and smell are properties. **(3)**

retain to hold **(45)**

rock a solid earth material. Rocks are made of minerals. **(3)**

salt water water with salt. Salt water is found in seas and the ocean. **(55)**

sand rocks that are smaller than gravel, but bigger than silt **(14)**

sand dune a hill of sand formed by wind **(36)**

silt rocks that are smaller than sand, but bigger than clay **(44)**

soil a mix of sand, silt, clay, gravel, pebbles, and humus **(44)**

solid matter that holds its own shape and always takes up the same amount of space **(63)**

texture the way something feels **(10)**

valley a low area between mountains **(26)**

volcano a place where lava, ash, and gases come out from the earth **(24)**

weathering when rocks break apart over time to become smaller and smaller **(18)**

wind moving air **(6)**